ECLIPSE 1919

AND

THE GENERAL RELATIVITY THEORY

GATOT SOEDARTO

Gatot Soedarto

Copyright © 2014 Gatot Soedarto.

All rights reserved.Including the right to reproduce this book or portions thereof, in any form. No part of this text may be reproduced in any form without the express written permission of the author

ISBN-13: 978-1500720612

ISBN-10: 1500720615

Printed in USA by CreateSpace Independent Book Publisher

Last update, August 29, 2017

Eclipse 1919 and the General Relativity Theory

Dedication

I dedicate this book for Nanda-Caca, Salsa-Sherly, Satriya-Ayu, and Abel Guruf Jr.

Contents

Preface	6
1. Einstein's Theory of Relativity.	9
2. Tests of General Relativity	13
3. Deflection of Light by Earth's Atmosphere	25
4. Fundamental Concepts of Astronomy	32
5. The Biggest Thing Einstein Was Wrong	39
6. Gravity Probe B and GPS	54
7. Conclusions	66
References	
About Author	

Preface

In the general theory of relativity, Einstein concluded that the light just as other material objects, moved in curve if gravity field of an object was massive. Einstein suggested that his hypothesis could be tested to observe the crossing track of the star light at gravity field of the sun. Due to the fact that the stars are not visible at day time, there is only one chance when the sun and the stars can be seen together at the sky, and that is the time when there is a solar eclipse.

Einstein suggested that the photo taken to the stars at the time when the sun was dark during the solar eclipse was compared to the photo of the same stars taken at other time / at night time. According to his hypothesis, the star light visible around the sun would be bent inwards, toward the sun at the time when passing through the gravity field of the sun, so that the picture of the stars would be visible for the observers on earth shifting outwards from their actual position in the sky. Einstein calculated the level of their deviation and predicted that for the stars observed being the closest to the sun, their deviation was about 1.75 seconds of arc.

Deviation or bending of star light was predicted by Einstein 1.75 seconds of arc without taking into account the altitude of star during eclipse, that is can not happens is astronomy; because deviation of star light is always varies depending on the altitude of star. It seems clear that Einstein had no idea on the basic of astronomy. This is actually a fatal mistake of Einstein. Aside a fatal mistake, Einstein suggests observations is made from the earth, it is also incorrect. When the observation is made from the earth, the result is always bending of light by earth's atmosphere, not bending of light by the sun.

This book discusses what's wrong with Einstein's hypothesis of general relativity, testing via eclipse, and the result of 1919 eclipse experiment. It has been known that Einstein never received a Nobel Prize for relativity since then. He was awarded a Nobel Prize of physics in 1921 for his work photoelectric effect. Why didn't the Nobel Committee in the year 1921 taking into account his theory of gravity?

Until now, for more than 90 years, the statement of Nobel Committee in the year 1921 is still valid:"Without taking into account the value that will be accorded your relativity and gravitation theories after these are confirmed in the future".

When you ask me, can the eclipse tell us if Einstein was right about general relativity? Unfortunately, I have to answer: No, that's impossible because Einstein's hypothesis is not valid. That's why there is no way to test intersection between general relativity and quantum theory through laboratory experiments.

Sidoarjo, August 29, 2017

Gatot Soedarto

1. Einstein's Theory of Relativity

There are two kinds of Einstein's theory of relativity. The first theory is special relativity announced in the year 1905 with the famous equation of $E = mc^2$, that describes the propagation of matter and light at high speeds.

The second theory, general relativity, was announced in 1916. This theory was born stimulated by the new fact just realized later on by Einstein that his theory on Special Relativity was found to be inconsistent with the gravity theory of Newton, declaring that the space objects pull to each others with the force whose magnitude is determined by distance between the said objects.

His hypothesis in theory on special relativity that the light velocity is the highest speed in this universe is in controversy with the gravity of Newton. The velocity of force attracting to each others among objects at the space, for example the attracting force of the moon causing the change in short time in the form of movement of sea tide on earth has the meaning that the gravity effect spreads at the boundless speed, not at the light velocity or lower.

Einstein's Hypothesis of Gravity

At his theory on general relativity, Einstein declared a new law on gravity, stating that gravity was not a force as commonly known at the Newton's gravity theory, but a part of inertia.. His gravity law illustrated the object behaviour at the gravity field, for instance the planets, not in the sense of *'the attracting forcer'* but only in the sense of the *crossing track* being taken.

Stephen Hawking write in his book ' A Brief History of Time ': "The theory also tells us that nothing can travel faster than light. The special theory of relativity was very successful in explaining that the speed of light appears the same to all observers (as shown by the Michelson-Morley experiment) and in describing what happens when things move at speeds close to the speed of light. However, it was inconsistent with the Newtonian theory of gravity, which said that objects attracted each other with a force that depended on the distance between them. This meant that if one moved one of the objects, the force on the other one would change instantaneously. Or in other gravitational effects should travel with infinite velocity, instead of at or below the speed of light, as the special theory of relativity required. Einstein made a number of unsuccessful attempts between 1908 and 1914 to find a theory of gravity that was consistent with special relativity. Finally, in 1915, he proposed what we now call the general theory of relativity. Einstein made the revolutionary suggestion that gravity is not a force like other forces, but is a consequence of the fact that space-time is not flat, as had been previously assumed: it is curved, or "warped," by the didistribution of mass and energy in it."(Stephen Hawking, A Brief History of Time, 1985).

For Einstein, gravity is a part of inertia. The movement of stars and planets originates from their inertia derivation, and the crossing track taken is determined by the space metrical nature, or more precisely the continuous space-time.

"Einstein's Law of Gravitation contains nothing about force. It describes the behaviour of objects in a gravitational field – the planets, for example – not in terms of ' attraction ' but simply in terms of the paths they follow. To Einstein, gravitation is simple part of inertia; the movement of the stars and the planets arise from their inherent inertia; and the courses they follow are determined by the metric properties of space – or, more properly speaking, the metric properties of the space-time continuum " (Lincoln Barnett, The Universe and Dr. Einstein, London, June 1949, page 72).

2. Tests of General Relativity

At his theory on general relativity, Einstein concluded that the light just as other material objects, moved in curve if gravity field of an object was massive. Einstein suggested that his hypothesis could be tested to observe the crossing track of the star light at gravity field of the sun. Due to the fact that the stars are not visible at day time, there is only one chance when the sun and the stars can be seen together at the sky, and that is the time when there is a *solar eclipse.*

He suggested that the photo taken to the stars at the time when the sun was dark during the solar eclipse was compared to the photo of the same stars taken at other time / at night time. According to his hypothesis, the star light visible around the sun would be bent inwards, toward the sun at the time when passing through the gravity field of the sun, so that the picture of the stars would be visible for the observers on earth shifting outwards from their actual position in the sky. Einstein calculated the level of their deviation and predicted that for the stars observed being the closest to the sun, their deviation was about 1.75 arc second.

The tests or proving method of a theory as suggested by Einstein was recorded at the book 'The Universe and Dr. Einstein' written by Lincoln Barnett, published for the first time in London in June 1949. The Preface of this book was written by Albert Einstein himself.

"From these purely theoretical considerations Einstein concluded that light, like any material object, travels in a curve when passing through the gravitational field of a massive body. He suggested that his theory could be put to test by observing the path of starlight in the gravitational field of thesun. Since the stars are invisible by day, there is only one occasion when sun and stars can be seen together in the sky, and that is during an eclipse.

Einstein proposed therefore, that photographs be taken of the stars immediately bordering the darkened face of the sun during an eclipse and compared with photographs of those same stars made at another time. According to his theory, the light from the stars surrounding the sun should be bent inward, toward the sun, in traversing the sun's gravitational field; hence the images of these stars should appear to observer on earth to be shifted outward from their usual positions in the sky.

Einstein calculated the degree of deflection that should be observed and predicted that for the stars closest to the sun the deviation would be about 1.75 seconds of an arc.

Since he staked his whole General Theory of Relativity on this test, men of science throughout the world anxiously awaited the findings of expeditions which journeyed to equatorial regions to photograph the eclipse of May 29, 1919. When their pictures were developed and examined, the deflection of the starlight in the gravitational field of the sun was found to average 1.64 seconds – a figure as close to perfect agreement with Einstein's prediction as the accuracy of instruments allowed." (Lincoln Barnett, The Universe and Dr.Einstein, London, Victor Gollanez LTD, First Published June 1949, Preface by Albert Einstein, page 78-79).

Actually Albert Einstein proposed three tests of the general relativity theory:

1.the perihelion precession of Mercury's orbit.

2. the deflection of light by the Sun.

3. the gravitational redshift of light

In the letter to the London Times on November 28, 1919, he described the theory of relativity and thanked his English colleagues for their understanding and testing of his work. He also mentioned three classical tests with comments:

"The chief attraction of the theory lies in its logical completeness. If a single one of the conclusions drawn from it proves wrong, it must be given up; to modify it without destroying the whole structure seems to be impossible." (http://en.wikipedia.org)

Reviews on the tests via solar eclipse

The tests / proving method as suggested by Einstein was carried out by a team of English scientists led by Arthur Stanley Eddington. Based on data from the Royal Astronomical Society, Arthur Eddington aimed to the group of *Hyade stars* from the city of Oxford in England at the nights in the months of January and February 1919. After that, Eddington together with his team left for Principe Island in West of Africa, and aimed to the Hyade stars during the solar eclipse on the date of May 29, 1919 at the city of Roca Sundy.

In the month of May 1919 the weather in Principe was not favourable because it was cloudy, and so was the time before the eclipse. However, Eddington succeeded in taking the photo of the solar eclipse taking place for 6 minutes 30 seconds. And the calculation output on light deviation by Arthur Eddington was 1.62 second arc, close to the Einstein's calculation output of 1.75 second arc.

The proving method for hypothesis as suggested by Einstein as the theory founder should not be able to be carried out, considering the fact that in scientific exposure in astronomy, the instant observation applies. It means, all calculations to determine the *'true position'* and the *'apparent position'* of a certain star at the sky is only applicable at a certain time and at a certain place on which such observation is performed.

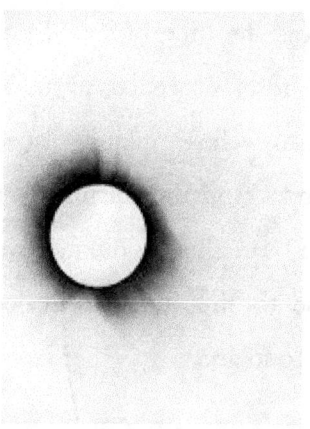

Figure 2.1: One of Eddington's photographs of the 1919 solar eclipse experiment, presented in his 1920 paper announcing its success (en.wikipedia.org)

The observation on a star conducted twice from the places with different geographical positions will result the different height/altitude and azimuth of the star. The altitude and azimuth of a star indicates the position of the star at the time when the observation is performed. The altitude and azimuth of a star changes every time due to the daily movement of the said space objects.

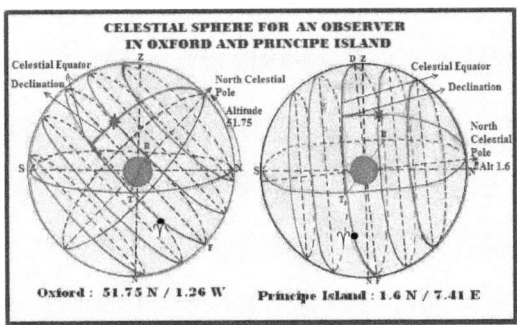

Figure 2.2: All calculations to determine the *true position* and the *apparent position* of a certain star at the sky is only applicable at a certain time and at a certain place on which such observation is performed.

Therefore, the proving method as conducted by Arthur Eddington, *should not be able to be performed.* Moreover, the observation / photo taking for the stars were performed twice with sufficiently long different interval of time.

Below is an illustration picture commonly used to explain 'light deviation' at the sun gravity field

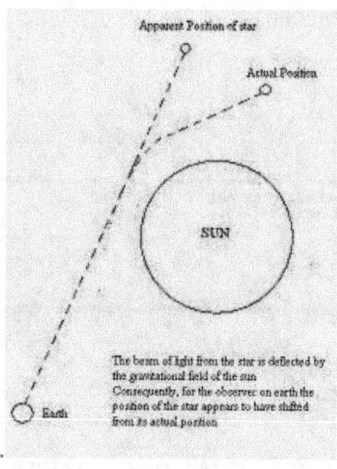

Figure 2.3: Bending of light by the Sun

In astronomy, the light deviation is something very common, and not caused by gravity field of a massive object, but it occurs due to the light refraction. Light refraction causes the light of all objects in the sky reaching the earth and seen by the observers, has been deviated by the media to pass through, including the light bending by the earth atmosphere.

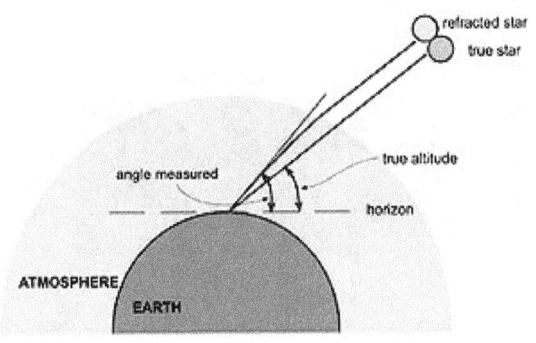

Figure 2.4: Refraction does make a star appear higher in the sky than it actually is (www.ngawhetu.com)

The magnitude of light deviation in astronomy is known as *"Astronomical Refraction"*, and can be calculated by using the *Snell's Law*.

The tests or proving method on the theory of general relativity as requested by its founder, Albert Einstein, is unjustifiable from scientific point of view of the astronomy. In addition, a hypothesis stating that the light is bent by gravity of massive object ignores the existence of light refraction (Snell's Law).

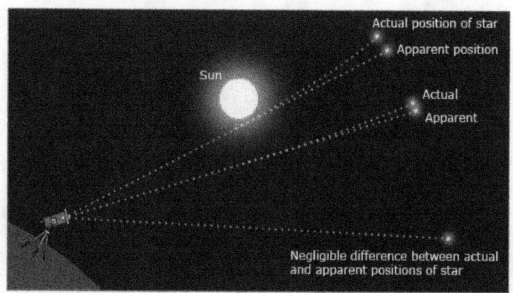

Figure 2.5: If Einstein's theory of relativity was correct, then the light from stars that passed closest to the sun would show the greatest degree of "bending." (undsci.berkeley.edu)

It is really hard to understand that the tests / proving method was conducted by a team led by Arthur Eddington. Therefore, output of the said proving performed by Arthur Eddington must be rejected. Moreover later on it was found out that based on the data from RAS (Royal Astronomical Society) that in the year 1919 RAS had sent two teams of expedition, namely Arthur

Eddington and Edwin Cottingham to Principe Island in West of Africa, Andrew Crommelin and Charles Davidson to Sobral in North East of Brazil.

The Celestial Sphere

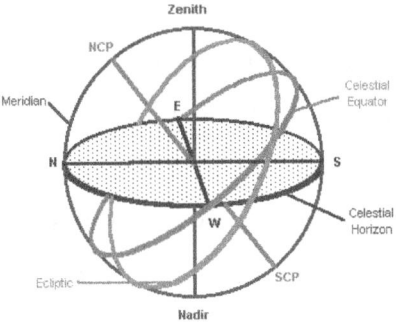

Figure 2.6: The Celestial Sphere is only applicable at a certain time and at a certain place on which such observation is performed.

Calculation output of Arthur Eddington's team was 1.62 second arc, while the calculation output of Andrew Crommelin was 0.93 second arc. This very vivid difference was ignored as if the expedition team performing the measuring from

Sobral never existed.

"Einstein's prediction of light deflection could not be tested immediately in 1915, because the First World War was in progress, and it was not until 1919 that a British expedition, observing an eclipse from West Africa, showed that light was indeed deflected by the sun, just as predicted by the theory. This proof of a German theory by British scientists was hailed as a great act of reconciliation between the two countries after the war. It is ionic, therefore, that later examination of the photographs taken on that expedition showed the errors were as great as the effect they were trying to measure. *Their measurement had been sheer luck, or a case of knowing the result they wanted to get, not an uncommon occurrence in science.* " (Stephen Hawking, A Brief History of Time, 1985).

3.Deflection of Light by Earth's Atmosphere

Light is naturally available in our surrounding, either in day time or at night. Such light may derive from the natural or artificial resources. When we see an object located far from our standing place, we think that what we see is its actual appearance. We are frequently unaware that what we see is actually not the real appearance of such object.

For example, once upon a time when we are at a beach and admiring the beauty of nature before the sun set. The sun looks moving down slowly, and in a certain time the lower part of the sun touches the edge of the sky or horizon. The panorama is so beautiful. However, when we see that beautiful panorama we are not aware that the actual sun has already fallen under the horizon. So, what we see is not the actual sun, but the *apparent sun*, or the sun on its *'apparent position'*. Even, the horizon or the sky edge that we see is not the *actual sky edge*, but the *illusive sky edge*.

Light refraction

Such phenomena is caused by the occurrence of light refraction reaching to our eyes. The light refraction causing the presence of apparent sun is called *'astronomical light curve'* or **'***Astronomical Refraction'*, whereas the thing causing the presence of the illusive sky edge is called *'earthy light curve'* or *"Terrestrial Refraction"*. The said terrestrial refraction causes the phenomena of *'Mirage'*, **a**nd mirage is not an optical illusion, but an actual physical phenomena.

So is at night time when the sky is clear and we can see and admire the stars sprinkle and spread at the sky. All those space objects are not in their true conditions, but on *their apparent position*, and all of them are caused by the astronomical refraction.

From the above explanation a question arises:

Cannot we ever see in our bold eyes a star in the sky on its true position condition? Such a chance is available, though limited, and will be found at the following discussion.

The ray curve occurs because the lights of an object reaching to our eyes / observers are not transmitted in the form of straight lines, but deviated by a medium all along its track, including the deviation by the earth atmosphere. The ray curve is an angle occurring between *the apparent position direction* and *the true position direction* of the said object.

The light of the stars in the sky reaches the earth passing through a very long distance so far away, and has already passed through various kinds of medium respectively having different densities. The classical scientists such as *Aristotle, Rene Descartes, Sir Isaac Newton* and others believed that the light of the stars reaching us on earth crept spreading through a medium the so-called luminiferous ether.

However various kinds of experiments had been made, among other was an experiment conducted by the American Scientists Michelson and Morrey 19th century, and all of those experiments failed to detect the presence of luminiferous ether, so that the ether is deemed non-existent. There is a possibility that luminiferous ether truly exists, but it cannot be proven.

Air density and Snell's Law

It is quite clear that the lights of objects in the sky reaching the earth have passed through layers of the terrestrial atmosphere, known as having different *air density*. Closer to the earth surface, the air is denser compared to the density of the air layer above it. The density is getting looser or weaker when it is getting higher.

The Snell's Law on light refraction declares that if a ray of lights passes through from one medium to the others with different densities, such ray of lights will be reflected. The magnitude of refraction angle depends of density of its medium. For example, a ray of lights is passed though water, the said ray of lights will be reflected closer to the normal.

On the picture below it is illustrated that the normal line is N-N'. The light passes through from A to B, and the light track forms the angle ABN. The ABN angle is called *'The Coming Angle'* or *"Angle of Incidence"*.

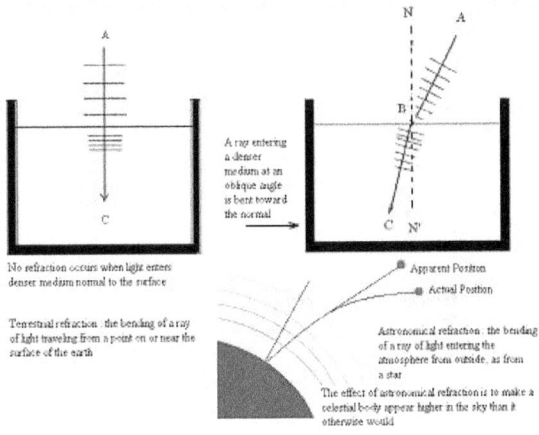

Figure 3.1: Refraction of light

In water, direction of light track is reflected close to the normal line, namely the BC direction, and forms the 'CBN Angle'. This CBN Angle is called the *'Refraction Angle'* or the *"Angle of Refraction"*. And the sinus of the Incidence Angle and sinus of the Refraction Angle have a fixed ratio. Such a ratio is called *"Index of Refraction"*.

A ray of lights is not reflected if its track is at the same direction with the normal. This statement answers the above question, that the change and the only chance to see a star on its true position is at the moment when the said star is precisely located straight above our head as an observer, or exactly at the Zenith point.

At the above picture, the difference between *air density* and *water density* is sufficiently big or in a sudden, therefore the light track in the air and in water looks like a broken line. It is completely different from the light track at the earth atmosphere. The air density at the layers of earth atmosphere changes gradually and regularly. This causes the light refraction in the form of a curve,and the effect of such curve, the apparent position of a star will always look higher than its true position.

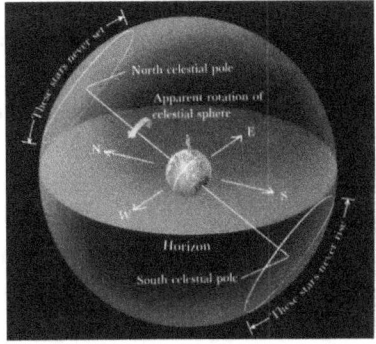

Figure 3.2: This figure illustrates that, depending on your latitude, some stars will be "circumpolar" and will never set. Remember: your latitude = the *altitude* of the north celestial pole. (astro.wsu.edu/worthey/astro/)

A curve of ray or light deflection is also known at the theory of Einstein, namely a deflection of light when passing through a gravity field of a massive object. According to this theory, when the light of a star passes through a gravity field of the sun, the said light will be deflected inwards, so that there will also be the *'apparent position'* and the *'true position'* of a star.

4. Fundamental Concepts of Astronomy

The observation on the stars in the sky at night give an idea, that all the stars are located at a surface of the space perfectly round circle. In astronomy, perfectly round circle is called the Celestial Sphere. And we as observers are in the center of the celestial sphere. The celestial sphere is an imaginary of a dome or a hemispherical screen. The celestial sphere is a practical tool for spherical astronomy, allowing observers to plot positions of objects in the sky when their distances are unknown or unimportant.

If we want to determine position of a point on the celestial sphere, it was first envisaged the existence of a horizontal front through the eye of the observer. The front imagination through the eyes of an observer is a special front on the earth, because of this imaginary front parallel to the surface sphere of arbitrarily large radius, concentric with earth. All objects in the observer's sky can be thought of as projected upon the inside surface of the celestial sphere, as if it were the underside of the sea. In astronomy, this front is called Horizon.

Eclipse 1919 and the General Relativity Theory

The Horizontal Coordinate System

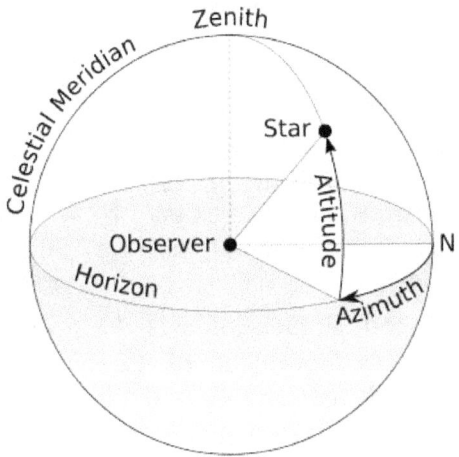

Figure 4.1: Azimuth is measured from the North point (sometimes from the South point) of the horizon around to the east; altitude is the angle above the horizon. (wikipedia.org)

Altitude and Azimuth of Star

- Altitude : is the angle between the line of the horizon and direction of the star seeing by an observer. Alternatively, some arc of the circle straight through the center of celestial bodies, is calculated from the front of the horizon to the star.

- Distance of culmination of a star, is the angle between the line direction of a star with the normal line of an observer. Alternatively, some arc of the circle straight through the center of celestial body, calculated from Zenith to the Star.

- Azimuth : is the angle between the celestial meridian with the star meridian . Alternatively, some arc of the horizon (the earth's equator), calculated from points north or south to the point of the circle sit up straight through the center of the celestial body.

Azimuth and Altitude of stars change at any time due to the daily movements of the stars. Therefore, writing azimuth and altitude of the stars must be included local time (to mention hours, minutes, and seconds), and the location of an observer (mention latitude and longitude), as well as the height of an observer calculated from the surface of the sea.

The observation on a star conducted by two people at the same time from different places, the results are different. Observing on a star by two people at the same place and time, but different heights of observers, the result is also different. The difference is due to the factor of 'astronomical refraction' and 'terrestrial refraction'. Therefore, Celestial Sphere is only applicable at a certain time and at a certain place on which such observation is performed.

Eclipse 1919 and the General Relativity Theory

Figure 4.2: Using sextant at sea -Image from practicalboating.com

Aside from Horizontal System in Celestial Coordinate Systems we know Equatorial system, Ecliptic system, and Galactic system.

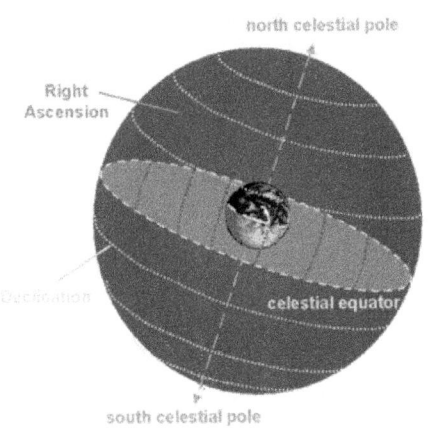

Figure 4.3: The **equatorial coordinate system-Image from astronomy.swin.edu.au**

The **equatorial coordinate system** is basically the projection of the latitude and longitude coordinate system on Earth, onto the celestial sphere.

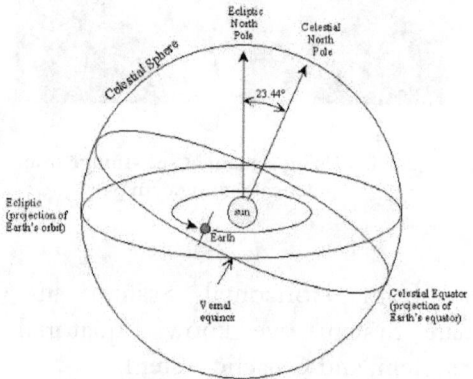

Figure 4.4: The Ecliptic Coordinate System-Image from coolcosmos.ipac.caltech.edu

The most important thing related to the proving method of general relativity theory is the celestial sphere, each place or an observer on Earth has its own celestial sphere. Examples:

Eclipse 1919 and the General Relativity Theory

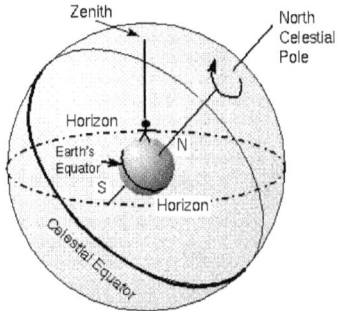

The celestial sphere for an observer in Seattle.
The angle between the zenith and the NCP = the
angle between the celestial equator and the horizon.
That angle = 90° − observer's latitude.

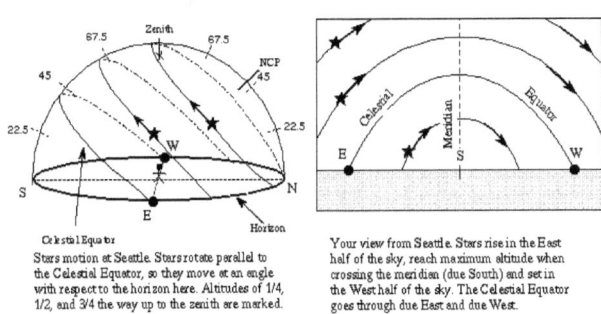

Stars motion at Seattle. Stars rotate parallel to
the Celestial Equator, so they move at an angle
with respect to the horizon here. Altitudes of 1/4,
1/2, and 3/4 the way up to the zenith are marked.

Your view from Seattle. Stars rise in the East
half of the sky, reach maximum altitude when
crossing the meridian (due South) and set in
the West half of the sky. The Celestial Equator
goes through due East and due West.

Figure 4.5: An example the celestial sphere for an observer at Seatle-Image from abyss.uoregon.edu.

Another example the celestial sphere for an observer at Mountain View and Los Angeles.

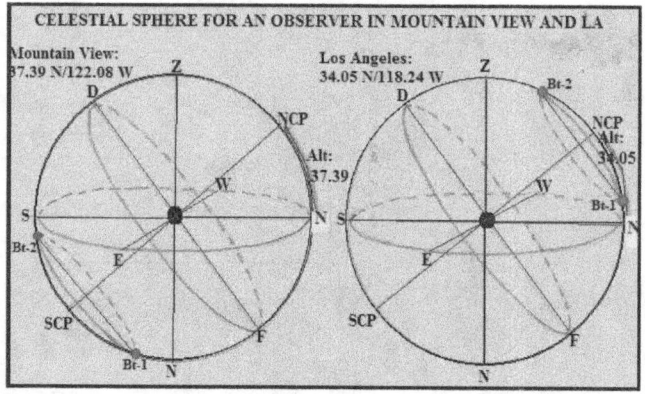

Figure 4.6:

Again, celestial sphere is only applicable at a certain time and at a certain place.

5. The Biggest Thing Einstein Was Wrong

Basic of astronomy.

When taking observe for finding the angle of bending of star light in the sky, knowing the altitude of star is just as important as measuring the time of moving celestial bodies. We must know the altitude of star quite exactly if observations are to be of any use. The reason for this is that the calculations of deviation all depending on the altitude of star.

Since the 18th century, astronomers had published the Nautical Almanac; can be used to facilitate the calculation. Nautical Almanac provides some of correction table, including correction for apparent altitude of Sun, Stars, and Planets This is called the astronomical refraction or refraction. Correction for height of eye an observer is called as DIP correction. This is actually the terrestrial refraction. That is recognized by the entire navigator in the world for a long time ago. Therefore, it is difficult to understand that the Nautical Almanac was not used in the 1919 eclipse experiment; whereas the hypothesis and testing via eclipse is closely related to the calculations in the Nautical Almanac of 1919.

Einstein's hypothesis and test via eclipse.

If you study the general relativity you will find that Einstein used the thought of experiment. The most famously Einstein's thought experiment is Einstein's elevator. From his thought experiment Einstein had found the Equivalence Principle. The original equivalence principle, as described by Einstein, concluded that free-fall and inertial motion were physically equivalent. Einstein said no-just as Galileo imagined the indistinguishability between a person inside a smooth-sailing ship (confined without windows) and a person on land, Einstein realised that the effects of acceleration and gravity were indistinguishable too. This is called the equivalence principle. A direct consequence of the equivalence principle is that light should be deflected or bent by gravity.

Let's look Einstein's idea on his elevator.

A cable is attached to the roof of the elevator;

some supernatural force begins reeling in the cable; and the elevator travels "upward" with constant acceleration, i.e. progressively faster and faster. Again the men in the car have no idea where they are, and again they perform experiments to evaluate their situation. This time they notice that their feet press solidly against the floor come up beneath them.

If they release objects from their hands, the objects appear to "fall". If they toss object in a horizontal direction they do not move uniformly in a straight line, but describe a parabolic curve with respect to the floor.

And so the scientist, who have no idea that their windowless car actually is climbing through interstellar space, conclude that they are situated in quite ordinary circumstances in a stationary room rigidly attached to the earth and affected in normal measure by the force of gravity. There is no way for them to tell whether they are at rest in a gravitational field or ascending with constant acceleration through outer space where there is no gravity at all. (Lincoln Barnett, The Universe and Dr.Eintein, page 69)

Figure 5.1: The objects appear to fall

From these fanciful occurrences Einstein drew a conclusion of great theoretical importance. To physicist it is known as the Principle of Equivalence of Gravitation and Inertia. It simply states that there is no way to distinguish the motion produced by inertial forces (acceleration, recoil, centrifugal force, etc) from motion produce by gravitational force.

The Einstein equivalence principle is the heart and soul of gravitational theory, for it is possible to argue convincingly that if Equivalence Principle valid, then gravitation must be a "curved spacetime" phenomenon, in other words, the effects of gravity must be equivalent to the effects of living in a curved spacetime.

But, these thought experiment can also be used to proves that the velocity of light is not constant; and the velocity of light is not the same speed as speed of light (300.000 Km/Seconds). Let's look the illustration below.

Figure 5.2: The light beam comes from above the elevator.

Imagine that the elevator still travels upward with constant acceleration, and a light beam comes from above the elevator. If the observer within the elevator are equipped with sufficiently delicate instruments of measurement, they will be able to compute the speed of light beam. The results show that the speed of light beam is faster than the speed of light 'c' 300.000 Km/Seconds (because the elevator travels upward!).

The observer drew a conclusion of great theoretical importance that the speed of light is not constant or is not the same speed as the speed of light 'c'. Thus, it's proves Einstein's special relativity really is false.

What's we learned from these thought experiment? That's show the fact, the thought experiment can be used to obtain the result they wanted to get. Einstein never proves his theories with the real experiments.

Also, Einstein relied on his mathematical equations: the Einstein Field Equation of Gravitation (EFE). He (and most of his followers) had not realized that EFE actually is similar with the thought experiment.

If you study Einstein Field Equation of Gravitation, you will find EFE is based on assumptions, thought experiment, and Riemann geometry. One of the basic topics in Riemannian geometry is the study of curved surfaces in general.

Riemann geometry also study higher dimensional spaces. But, there are no practical applications of Riemann geometry in astronomy. Riemann did not take an interest in the space of the astronomers. Questions about the global properties of space he cut short as "idle questions."

$$R_{\mu\nu} - \frac{1}{2} R\, g_{\mu\nu} + \Lambda\, g_{\mu\nu} = \frac{8\pi G}{c^4} T_{\mu\nu}$$

Figure 5.3: Curvature of spacetime = density and flux of energy and momentum.

The left-hand side of that equation is a matrix of numbers (curvature of spacetime). Mathematically, spacetime is a manifold. The right-hand side is a matrix of quantum operators, each of which has an expectation value. This, at some level, makes no sense.

Mathematically, spacetime is a manifold. But, there is no manifold in the basic of astronomy. Albert Einstein really had no idea on the basic of astronomy. It's like a people have no experience in electronics, then try to modify a sound system, of course, the resulting sound is becoming discordant.

Without having experience on the basic of astronomy; Einstein drew a conclusion the hypothesis of general relativity and proposed test via solar eclipse that are closely related to astronomy.

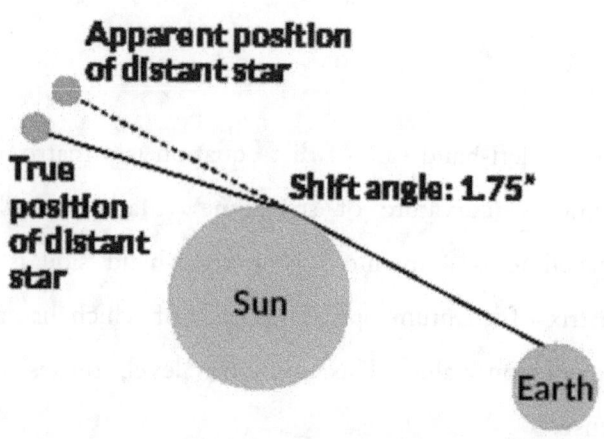

Figure 5.4:Bending of light-Image from sciencenews

According to Einstein, the star light visible around the sun would be bent inwards, toward the sun at the time when passing through the gravity field of the sun. Einstein calculated the level of their deviation and predicted that for the stars observed being the closest to the Sun, their deviation was about 1.75 seconds of arc.

For more than 100 years all physicists and astrophysicists are very familiar with the illustration on the Figure 5.4 above; but, did they realize that the above illustration has no meaning whatsoever? The above illustration shows Einstein has been failure to understand the basic of astronomy.

What is the reason? This prediction is not meaningful in scientifically of astronomy when it is not explained the altitude of the star/Sun. The important things to be noted, the amount of 1.75 seconds of arc without taking into account the altitude of the star as the object of observation. This is a fatal mistake; because the deviation of starlight will always varies depending on the altitude of the object of observation.

Thus, Einstein's hypothesis of general relativity is not valid. Something weird and magical when experimental test's of general relativity; testing via eclipse (1919, 1922, 1929, 1936, 1947,1952, and 1973) the result is declared general relativity was correct. This is very embarrassing to the world of science. They are not aware of making themselves look so foolish. Lack of knowledge on the basic of astronomy.

They also declared, that in modern times, some of the most precise measurements of light-bending have come from radio observations of distant galaxies; known as measurement using the Very Long Baseline interferometry (VLBI) to obtain results that supported Einstein's predictions.

The Nature website tells us about new measurement of solar gravitational deflection of radio signals using VLBI.

"RADIO observations using very-long-baseline interferometry (VLBI) can measure the deflection of electromagnetic radiation by the Sun's gravitational field with an accuracy of better than 1 milliarcsecond, and can thus be used to test General Relativity. For an object at an angle a from the centre of the Sun, the expected deflection is1 $(1 + \text{gamma})\,(Ms/re)((1 + \cos \text{alpha})/(1-\cos \text{alpha}))1/2$, where Ms is the mass of the Sun in geometrized units2 (1.477 times 105 cm), re is the distance from the Earth to the Sun in cm, and y is a parameter whose value is 1 if General Relativity is correct but which takes on different values in other theories of gravity. For gamma = 1, the deflection is 1,750 mas at the Sun's limb, 4 mas at alpha =90° and 0 at alpha = 180°. Our analysis of ten years of VLBI data, including observations of objects in the range 2.5° < a< 178°, yields an estimate gamma = 1.0002 with a formal standard error of 0.00096 and an estimated standard error of 0.002. This determination is comparable in accuracy and in good agreement with the determination from Mars–Viking time-delay measurements"

We can read from the abstract above 'Our analysis of ten years of VLBI data ..." It means collecting data within ten years, then comparing and analyzing data. This method is not justified in scientifically of astronomy. To measure the angle of deviation of starlight, in astronomy applies direct observation and instantaneous.

They did not hesitate to say: "One century after its formulation, Einstein's general relativity has made remarkable predictions and turned out to be compatible with all experimental tests."; or said that special and general theory of relativity are incredibly well tested and very accurate theories.

They did nor realize that 'the primary problem' of general relativity is not experimental test but hypothesis and Einstein proposed test via eclipse.

The leading science writer, Lincoln Barnett, tells us in his book:

"From these purely theoretical considerations Einstein concluded that light, like any material object, travels in a curve when passing through the gravitational field of a massive body. He suggested that his theory could be put to test by observing the path of starlight in the gravitational field of the Sun. Since the stars are invisible by day, there is only one occasion when Sun and stars can be seen together in the sky, and that is during an eclipse.

Eclipse 1919 and the General Relativity Theory

Einstein proposed therefore, that photographs be taken of the stars immediately bordering the darkened face of the sun during an eclipse and compared with photographs of those same stars made at another time. According to his theory, the light from the stars surrounding the Sun should be bent inward, toward the Sun, in traversing the Sun's gravitational field; hence the images of these stars should appear to observer on earth to be shifted outward from their usual positions in the sky.

Einstein calculated the degree of deflection that should be observed and predicted that for the stars closest to the Sun the deviation would be about 1.75 seconds of an arc.Since he staked his whole General Theory of Relativity on this test, men of science throughout the world anxiously awaited the findings of expeditions which journeyed to equatorial regions to photograph the eclipse of May 29, 1919. When their pictures were developed and examined, the deflection of the starlight in the gravitational field of the sun was found to average 1.64 seconds—a figure as close to perfect agreement with Einstein's prediction as the accuracy of instruments allowed."(Lincoln Barnett, The Universe and Dr.Einstein, page 78)

Explanation:

1.Einstein predicts of 1.75 sec.arc without taking into account the altitude of the star as the object of observation. This is a fatal mistake because this prediction has no scientific meaning in astronomy. The important thing to be note, in astronomy, deviation or bending of starlight will always vary depending on the altitude of the object of observation. In this case, hypothesis Einstein is not valid. Einstein hypothesis of GR doesn't meet requirements of scientific method.

2."Einstein proposed therefore, that photographs be taken of the stars immediately bordering the darkened face of the sun during an eclipse and compared with photographs of those same stars made at another time. According to his theory, the light from the stars surrounding the Sun should be bent inward, toward the Sun, in traversing the Sun's gravitational field; hence the images of these stars should appear to observer on earth to be shifted outward from their usual positions in the sky."

This sub-paragraph shows that he wants measuring deflection of light by the Sun; but he proposed test measuring deflection of light by Earth's atmosphere; he had not realized about that. Ironically, this test is not scientifically correct and deeply wrong.

a. When observation is made from Earth, not from outer space, the result is always deflection of starlight by Earth's atmosphere.

b. Calculating the angle of deflection of light

(bending, deviation), in astronomy applies direct observation and instantaneous. "compared with photographs of those same stars made at another time" is not scietifically correct.

That's why until now, for more than 90 years, the statement of Nobel Committee in the year 1921 is still valid:"*Without taking into account the value that will be accorded your relativity and gravitation theories after these are confirmed in the future*".

6. Gravity Probe B and GPS

Gravity Probe B experiment

Gravity Probe B (GP-B) is a NASA physics mission to experimentally investigate Albert Einstein's 1916 general theory of relativity. GB-B uses four spherical gyroscopes and a telescope, housed in a satellite orbiting 642 km (400 mi) above the Earth, to measure in a new way, and with unprecedented accuracy, two extraordinary effects predicted by the general theory of relativity, the second having never before been directly measured:

1. The geodetic effect, the amount by which the Earth warps the local spacetime in which it resides.

2. The frame-dragging effect, the amount by which the rotating Earth drags its local spacetime around with it.

According to NASA, the Gravity Probe B gyroscopes are the most perfect spheres ever made by humans. If these ping pong-sized balls of fused quartz and silicon were the size of the Earth, the elevation of the entire surface would vary by no more than 12 feet.

We interested to finding the relationships between spacetime and atmospheric medium, as we know that in general relativity Einstein had ignored the atmospheric medium around a massive body in the sky.

Earth's gravity pulls all the objects in atmosphere toward the Earth, and all the objects follow the rotation of the Earth. In the same way the satellite Gravity Probe B has orbit at the altitude 400 miles (642 km) above the Earth, remain in orbit at thermosphere/exosphere, and doesn't escape to space, it's because Earth's gravitational force. That is a facts, no doubt about it.

Something weird and illogical if someone had been ignored the effects of Earth's gravitational force, and then tries to prove 'gravity is nothing about force' in the form of the geodetic effect and the frame-dragging effect.

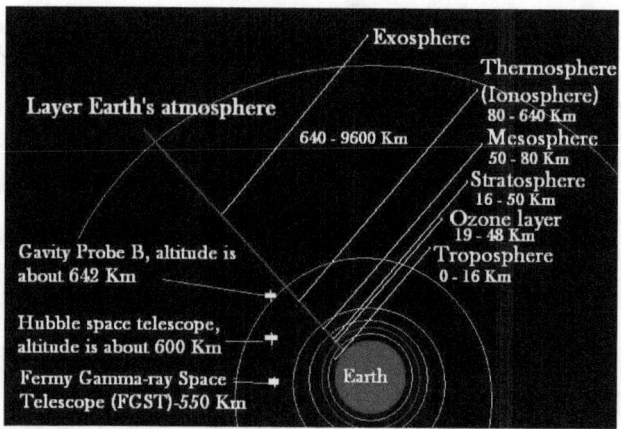

Figure 6.1: Gravity Probe B orbits in the Earth's atmosphere, the altitude is about 642 Km

The above figure shows Gravity Probe B orbits in the Earth's atmosphere, even though; the Earth's atmosphere is not spacetime. Thus, in this case, it can be seen as a false equivalence-describing the Earth's atmosphere is the same as spacetime, when in fact there is not the same.

How could to detect the geodetic effect and the frame-dragging effect in the Earth's atmosphere, although the Gravity Probe B gyroscopes are the most perfect spheres ever made by humans? If we assumed spacetime is the same with atmospheric medium, so we must change Einstein Field Equation of Gravitation (EFE). As we know, the meaning of EFE is: curvature of spacetime = density and flux of energy and momentum. Then, if the Gravity Probe B experiment was correct, we must change: the atmospheric medium = density and flux of energy and momentum.

Miles Mathis wrote in his article entitle Gravity Probe B and space-time:

In a nutshell, what the Gravity Probe experiment did is measure the tilt of little gyroscopes. If the tilt is zero, no curvature of space-time. If the tilt is not zero, we are supposed to have proof of curvature. The gyroscope tilts because space is curved.

The primary problem is that there is absolutely no effort in this experiment to consider, mention, or try to block the main cause of that tilt. It is simply assumed that any non-zero outcome is proof positive of their theory and that any tilt that does not match their needed numbers is only an anomaly or "observation" that can be explained away later.

That is horrible science, no matter how you look at it.

Gravity Probe B project that was given a grade of F in a NASA review in 2008 by a group of senior advisors, and denied any further funding since, "the reduction in noise needed to test rigorously for a deviation from general relativity is so large that any effort ultimately detected by this experiment will have to overcome considerable (and in our opinion, well justified) scepticism in the scientific community". They continue:

"The noisy data meant that GPB could not measure the effects as precisely as astronomers had by firing laser beams at mirrors left of the Moon by the Apollo astronauts."

Relativity and Global Positioning System (GPS)

Einstein's gravity was totally wrong. General relativity is not valid. What can general relativity do to GPS? Hypothesis and test of general relativity are closely related to astronomy, especially celestial navigation. If you would like to make sure general relativity is not valid, physics training is needed; but more importantly (especially for general relativity) is celestial navigation training.

GPS measuring location, not measuring time. It's about time dilation of special and general relativity. In the case of special relativity: time delay of light; it's clear that light can be bent by layer of atmosphere. Thus, speed of light is not constant.

"Nothing can travel faster than the speed of light."

"Light always travels at the same speed."

Have you heard these statements before? They are often quoted as results of Einstein's theory of relativity. Unfortunately, these statements are somewhat misleading. Let's add a few words to them to clarify. "Nothing can travel faster than the speed of light in a vacuum." "Light in a vacuum always travels at the same speed." Those additional three words in a vacuum are very important.Light traveling through anything other than a perfect vacuum will scatter off

off whatever particles exist.

What can special relativity do to GPS? Nothing. Signals from GPS are not sent back from the receiver on earth to the GPS.

University of London Professor Herbert Dingle showed why Special Relativity will always conflict with logic, no matter when we first learn it. According to the theory, if two observers are equipped with clocks, and one moves in relation to the other, the moving clock runs slower than the non-moving clock. But the Relativity principle itself (an integral part of the theory) makes the claim that if one thing is moving in a straight line in relation to another, either one is entitled to be regarded as moving. It follows that if there are two clocks, A and B, and one of them is moved, clock A runs slower than B, and clock B runs slower than A. Which is absurd.

In the case of general relativity, we know and no doubt: general relativity is not valid. We know that they did not hesitate to say: "One century after its formulation, Einstein's general relativity has made remarkable predictions and turned out to be compatible with all experimental tests."; or said that "special and general theory of relativity are incredibly well tested and very accurate theories." But, actually, these statement is nonsense.

About testing general relatiity via eclipse experiment using optical telescope; if it was difficult in 1995 , to see details of 1–2 seconds of arc, how much more difficult was it in the in 1919–1973 eclipse experiments? The difficulty of performing precise measurements of optical starlight deflection during an eclipse can be seen from the results of 1919, 1922, 1929, 1947, 1952, 1973 experiments.

Testing general relativity using VLBI (Very-long-baseline interferometry). VLBI is a type of astronomical interferometry used in radio astronomy. In VLBI a signal from an astronomical radio source, such as a quasar, is collected at multiple radio telescopes on Earth. The important things must be note, the purpose of VLBI is collecting signal in the form of invisible light, not to measure the altitude of a star and bending of light in the form of visible light.

VLBI can not be use as a sextant in celestial navigation. What can general relativity do to GPS? Nothing.

General relativity predicts the clocks on the satellites appear to be ticking faster than identical clocks on the ground. Actually, this prediction can be explained without Einstein's theory.

Clocks at higher altitude tick faster than clocks on Earth's surface. It is not caused by gravity, but caused by air density of atmosphere. Closer to the Earth surface, the air is denser compared to the density of the air layer above it. The density is getting looser or weaker when it is getting higher. The effect is the same for ordinary clocks or atomic clocks. Moreover, atomic clocks are sensitive to the temperature changes and pressure in their orbit.

It is has been known in traveling on an airplane. At higher altitude the density of amosphere is getting looser or weaker, and less of friction on an airplane. Traveling in weaker density of atmosphere an airplane can move faster than in denser atmosphere.

There are no official statements

How accurate is GPS? According to website dedicated by the USA government to the GPS, the answer is it depends. GPS satellites broadcast their signals in space with a certain accuracy, but what you receive depends on additional factors, including satellite geometry, signal blockage, atmospheric conditions, and receiver design features/quality.

For example, GPS-enabled smart phones are typically accurate to within a 4.9 m (16 ft.) radius under open sky. However, their accuracy worsens near buildings, bridges, and trees.

From the website, we know that GPS does not put forward anything about Einstein's relativity. In other words, there are no official statements.

Explanation of GPS's special consultant.

In the 1990's, Van Flandern worked as a special consultant to the Global Positioning System (GPS), a set of satellites whose atomic clocks allow ground observers to determine their position to within about a foot. Van Flandern goes on to discuss GPS clocks, which are often cited as being proof positive of Einstein's relativity. It may surprise you, but the GPS system doesn't actually use Einstein's field equations.

In fact, this paper by the U.S. Naval Observatory tells us that, while incorporating Einstein's equations into the system may slightly improve accuracy, the system itself doesn't rely on them at all. To quote the opening line of the paper, "The Operational Control System (OCS) of the Global Positioning System (GPS) does not include the rigorous transformations between coordinate systems that Einstein's general theory of relativity would seem to require."

At high altitude, where the GPS clocks orbit the Earth, it is known that the clocks run roughly 46,000 nanoseconds (one-billionth of a second) a day faster than at ground level, because the gravitational field is thinner 20,000 kilometers above the Earth. The orbiting clocks also pass through that field at a rate of three kilometers per second—their orbital speed. For that reason, they tick 7,000 nanoseconds a day slower than stationary clocks.

To offset these two effects, the GPS engineers reset the clock rates, slowing them down before launch by 39,000 nanoseconds a day. They then proceed to tick in orbit at the same rate as ground clocks, and the system "works." Ground observers can indeed pin-point their position to a high degree of precision. In (Einstein) theory, however, it was expected that because the orbiting clocks all move rapidly and with varying speeds relative to any ground observer (who may be anywhere on the Earth's surface), and since in Einstein's theory the relevant speed is always speed relative to the observer, it was expected that continuously varying relativistic corrections would have to be made to clock rates.

This in turn would have introduced an unworkable complexity into the GPS. But these corrections were not made. Yet "the system manages to work, even though they use no relativistic corrections after launch," Van Flandern said. "They have basically blown off Einstein".

7. Conclusions

The biggest thing Einstein was wrong because he had no idea on the basic of astronomy: Einstein's hypothesis of general relativity actually can not be proven or tested in any way.

Aside his hypothesis, the two fatal mistakes of Einstein; he wants measuring deflection of light by the Sun; but he proposed test measuring deflection of light by Earth's atmosphere; he had not realized about that. Ironically, this test is not scientifically correct and deeply wrong.

The tests of General Relativity as suggested by Einstein : the photo taken to the stars at the time when the sun was dark during the solar eclipse was compared to the photo of the same stars taken "*at another time*'. *The words 'at another time' means* proposal suggested by Einstein is not at all scientific, the reasons are explained by the following pictures.

Figure 6.1: Two stars in the sky as seen from A.

Figure 6.2: The same two star as seen from B at another time

Figure 6.1 is two stars portrait taken by an observer from point A in the earth at a time, and Figure 6.2 is the same portrait of two star, taken by the same observer from point B in the earth '*at another time*' eg two months later. Point A located in Europe, while the point B located in the Africa continent, the two point has a big different of the latitude and longitude.

The two sightings of stars will always be different because point A and B has its own Celestial Sphere. The Celestial Sphere is only applicable at a certain time and at a certain place on which such observation is performed. Therefore, the two portraits can not be compared.

And again, the two portraits are the appearance of apparent position of the two stars, not the the true position. Therefore, from the two portraits can not be use to calculate the angular difference between true position and apparent position of stars that becomes the object of observation.

The tests / proving method on the theory on general relativity as requested by its founder, Albert Einstein, is unjustifiable from scientific point of view of the astronomy. In addition, a hypothesis stating that the light is bent by gravity of massive object ignores the existence of light refraction (Snell's Law).

If a scientist conveys a theory and at the same time shows its proving method, however after being tested by another scientist it is found out that his proposed proving method is proven to be unable to be performed due to not being scientific, then automatically such proposed theory prematurely falls by itself. And the proving cannot be carried out by other methods not as requested by the theory founder, since it is reasonably assumed that such proving is made based on belief.

Hypothesis and Einstein proposed test of general relativity are closely related to astronomy, especially celestial navigation. For understanding that hypothesis and the test are not valid, physics training is needed; but more importantly is celestial navigation training. Unfortunately, physicists and astrophysicists are not trained to become experts in the field of celestial navigation. The navigators around the world will be easily to recognize the fatal flaws of these hypotheses and test. Actually, general relativity can not be proven or tested in any way.

No doubt, the general relativity has been wrong since the beginning; all tests that says 'general relativity is correct' really are the case of 'knowing the result they wanted to get'.

That's why until now, for more than 90 years, the statement of Nobel Committee in the year 1921 is still valid:"Without taking into account the value that will be accorded your relativity and gravitation theories after these are confirmed in the future".

Einstein's hypothesis is not valid. That's why there is no way to test intersection between general relativity and quantum theory through laboratory experiments

Eclipse 1919 and the General Relativity Theory

References

1. **Bowditch**, American Practical Navigator, Volume I - II, Defense Mapping Agency Hydrographic / Topographic Center, 1984.

2. **Lincoln Barnett**, Universe and Dr.Einstein, London, June 1949.

3. **Stephen Hawking**, A Brief History of Time, 1985.

4. **NASA** Eclipse Web Site.

5. **ScienceCentric** Web Site.

6. **Tom van Flanderm**,The Speed of Gravity: Why Einstein Was Wrong and Newton Was Right, November 29, 2012.

7. **Miles Mathis**, Gravity Probe B and space-time, milesmathiscom:

8. http://en.wikipedia.org

9. http://astro.wsu.edu/worthey/astro

10. http://undsci.berkeley.edu

11. http://astronomy.swin.edu.au

12. http://coolcosmos.ipac.caltech.edu

13. http://www.rpi.edu/

About the Author

Capt (Ret) Gatot Soedarto, was born in Tuban, East Java, he graduated from The Indonesia Naval Academy. Former Chief of The Indonesia Coast Guard's Fleet, lecturer at The Indonesia Naval Academy, lecturer at The Indonesia Naval Staff and Command College, lecturer at The Indonesia Armed's Command School.

The books of his works among others are: Computer Engineering (1981), Prevention and Coping with Fire Hazards (1983), Preventing the Environmental Destruction from Fire Hazard (1985), Stress and How to Overcome it (1989), Sun Tzu and Naval Strategy (2012), Lessons of the Falklands War (2013), Rubik's Cube for beginners (2014),Einstein,Arthur Eddington dan Astronomi(2014).

Twitter: @GatotSoedarto
Facebook: Gatot S.Astari

www.ingramcontent.com/pod-product-compliance
Lightning Source LLC
Chambersburg PA
CBHW071302170526
45165CB00003B/1383